GOING SOLAR

THE HOMEOWNER'S HANDBOOK

LOURDES DIRDEN

THINK BOOKS

GOING SOLAR
The Homeowner's Handbook

Revised 2nd edition, Jun 2022
Revised 1st edition, Nov 2021

Cover Design by Raffi Antounian
Editing by Gretchen Pruett

Library of Congress Control Number: 2020924375

Print ISBN: 978-1-7346592-8-3
Ebook ISBN: 978-1-7346592-1-4
1st edition, Mar 2021

Think Books
www.thinkbooks.org

ACKNOWLEDGMENTS

To all the homeowners who shared their experiences, you are the reason this book was written. Thank you.

To my husband, family, and friends, thank you for your love, prayers, and support.

To Raffi, my cover designer, you are incredible. Thank you for your magnificent art work.

To Gretchen, my editor, you are amazing. Thank you for your outstanding work.

TABLE OF CONTENTS

PART 1

PART 2

PART 1

Too often, the opportunity knocks, but by the time you push back the chain, push back the bolt, unhook the two locks and shut off the burglar alarm, it's too late.

RITA COOLIDGE

CHAPTER 1

AN INTRODUCTION

How often misused words generate misleading thoughts.

HERBERT SPENCE

The sun has been feared and worshiped over the course of history. Many cultures from around the world associate the sun with their most important deity. The sun is the only natural source of light in the entire solar system, and without it, we would not exist.

Since ancient times, heat and light have been harnessed from the sun. The sun's heat was used to light fires with glass and mirrors during the 7th century B.C. In the 3rd century B.C., the Greeks and Romans harnessed solar power with mirrors to light torches for religious ceremonies. Coming into the 2nd century B.C., there was an interesting tale about the Greek scientist, Archimedes. He used the reflective properties of bronze shields to focus the sun's power and set fire to wooden ships from the Roman Empire. His reasoning was Syracuse had been under siege for 2 years. This tale has been told in Greece for centuries and, in 1973, the Greek navy recreated the experiment and successfully set a wooden boat 50 meters away on fire.

Today, solar energy has helped create solar-powered buildings, vehicles, and remote-controlled aircraft—to name just a few. It is a form of renewable energy and as long as the sun shines, this energy source can be found in many areas of the world.

In my solar industry career, I have had the pleasure of working with a team of professionals. I have worn many hats—namely customer care, solar monitoring specialist, system troubleshooter, to name a few. These different roles have had an informative impact on my life. In this line of work, I have been in constant communication with homeowners. There have been happy, satisfied customers, and also some who were frustrated and angry. These experiences awoke a different avenue of awareness that I had no idea existed.

I noticed a common theme among homeowners. Many did not conduct their own research prior to going solar. I discovered that a lot of them did not know what questions to ask. To top it off, there was not much help from some of the salespeople, who were independent contractors. They did not offer information that would have helped some of these homeowners make an informed decision.

A reputable solar provider offers information and steps back to give the homeowner time to digest the information. This will help them make an educated decision. Yet, this does not happen often. A lack of basic understanding is why there are so many misconceptions around solar energy today. I wrote this book to address this problem.

Going Solar The Homeowner's Handbook was created for anyone who wants to raise their energy awareness. It will provide you with what you need to know about the process of having solar panels installed in your home.

Although some of the information pertains to people from

the United States, people from different parts of the world will also benefit. You will learn tips, the right questions to ask, identify when someone is feeding you the wrong information, and on Chapter 13, International Resources, you will find information and links about going solar.

My hope is that you will feel empowered and realize that awareness is ongoing and learning is just the beginning.

CHAPTER 2
THINGS TO CONSIDER

No matter what people tell you, words and ideas can change the world.

ROBIN WILLIAMS

This chapter will cover the following topics:

- Is solar good for every home?
- Will solar work on my roof?
- Does shade make a difference?
- Is my home in a good location?
- Will I save money?
- Where will my solar panels be placed?

There are several things to think about before you make a decision to purchase a solar system for your home. Always do your homework. Talk to your friends and neighbors who have chosen solar and evaluate the information. Always be active and engaged when dealing with solar providers.

Is Solar Right for Your Home?

A solar system is not for every home. The energy required to power your home depends on your energy consumption. Making your home energy efficient before going solar is an essential first step. By decreasing your overall energy use before going solar, you could reduce the size of the solar system you need, potentially saving thousands of dollars. The weight of the solar panels also play a considerable role because the roof structure must be able to withstand it. It is also important to know how many people live in your home, and how many electrical devices are used. Know your situation by evaluating your home. Review the following steps:

- Consider how long you plan to stay in your home.

- Trim your energy usage as much as possible.

- Follow your electric bills and determine how many kilowatts of energy you use on an average day.

- Know your electricity usage. Start by reviewing your utility bill to see how much energy you used in the last year and what it cost.

- *Ask yourself the following*: Are you planning any changes that will affect your electricity use, such as buying an electric vehicle, planning an addition to your home, adding an appliance (air conditioner, freezer, etc.), or improving your energy's efficiency?

- Know your electricity rates.

- *Ask yourself the following:* 1) Do your electricity rates differ depending on the time of day? 2) Does your utility charge a fixed monthly fee based on your peak electricity usage (also called a demand charge)? 3) Will the utility company compensate you for any extra electricity your solar system produces beyond your need and, if so, at what rate?

- Make your home and appliances more energy efficient.

- *Do the following:* Check for efficiency upgrades. For more information, go to the United States Department of Energy, Office of Energy Efficiency & Renewable Energy at https://www.energy.gov/eere/why-energy-efficiency-upgrades.

The Roof Matters

The condition of your roof is absolutely essential. It is a critical component to consider before installing solar panels. Proper installation begins before the installation starts. You need to know if your roof needs replacing before the solar panels are installed.

I have come across homeowners who called to schedule the removal of their panels because their roof needed replacement. When I informed them that there would be a cost, their response was that they were not aware. It is very important to read the contract (Chapter 8, "Providers, Contracts & Warranties") and know what is included in panel removal and reinstallation.

Another important consideration is the roof warranty compared to the warranty on the solar panels. If you have solar panels that will last 20 years, and a roof that will last between 10 to 15 years, the cost of maintenance increases. It is best to figure out how long the roof will last to reduce the amount of effort, time, and money you will need to invest. Ideally, you want your roof to have a longer warranty than the solar panels.

The National Association of Home Builders (NAHB) conducted a survey in 2007. They found that, on average, slate, copper and tile roofs can last more than 50 years; wood shake roofs about 30 years, cement shingles about 25 years, and asphalt shingle/composition roofs about 20 years. Roof manufacturers offer warranties that last between 10 to 30 years.

Solar panels can be installed on nearly every type of roof. There may be some that require more effort and extra costs. Here are some important aspects to consider about your roof.

— Roof Age

- Determine how long it will last.

- If it is more than 10 years, you may want to consider repairs or replacement.

- If it is more than 15 years, you may want to replace it.

— Roof Type

Solar panels can be safely and effectively installed on almost any roof materials, however, the installation process may differ. There are some jurisdictions that do not allow solar panels on a wood shake roof because of a potential fire hazard. A few of the most commonly used roof materials are asphalt shingles, clay tiles, slate tiles, concrete, rubber, wood, and metal. For further information, ask your solar provider.

— Roof Condition

Before hopping on a ladder, you can spot some easy signs of age. For example, leaks or stains in the attic or walls. If your energy costs have gone up in recent years, it may be worth it to see if it is due to excessive ventilation. From the outside, you can check to see if there is any mildew or if there are dark spots on the roof.

Solar panels will reduce some of the wear and tear on your roof. They block rain and protect it from years of consistent sunlight; however, unless you plan to get photovoltaic shingles (also known as Building-integrated photovoltaics), solar panels are not a replacement for your roof.

Shading Limits Sunlight

The sun hits your solar panels for at least five hours each day. Yet, when your panels are covered in the shade, less power is produced—which causes lower efficiency than those in direct sunlight. Look around the immediate surroundings of your home to assess for shade. Trees, hills, and other buildings may block the sun, and will greatly affect the level of solar radiation reaching your panels. For example, in the city, tall buildings could dramatically limit the amount of sunlight that reaches your home.

For the homes that have yards, imagine the future of your lawn. If the roof is unobstructed, but you have planted leafy oaks around your property, you may run into trouble a few years down the road. Be prepared to trim your plants and trees to keep the panels clear. Common items that create shade include trees, chimneys, nearby buildings, electrical cables, heating and cooling equipment, pipes, skylights, and vents.

The Importance of Location

In the United States, the average household uses 900 kilowatts of energy each month. A 5 kilowatt solar system exposed to just four hours of sunlight a day will produce 600 kilowatts of energy per month.

In most areas, solar panels will receive more than four hours of sunlight a day and produce more energy than you actually need. The extra energy will give you solar credits as long as your electric company has the net metering program (Chapter 6, "Net Energy Metering").

One of the key elements that determines how much sun your solar panels will see over the course of the day is the direction the roof faces. Solar panels produce most power when they are pointed directly at the sun facing towards the equator.

In the northern hemisphere, solar panels are most effective when installed on south-facing roofs. Although, even if your roof does not face directly south, they can still produce plenty of electricity, but the performance may be less powerful.

If you are aiming to cover all of your electricity usage with solar, here are some solutions:

- Install a few more panels than you would otherwise need with a southern-facing system.

- Install ground-mounted solar or a carport installation.

Will Solar Energy Save Me Money?

A simple answer would be the best. Unfortunately, solar energy is a little more complicated. Saving money on solar depends on a few factors. A solar provider can assist you in figuring out if you will save money, but it would be a good idea if you knew the answers to some of these questions.

- How much sun hits my roof?

- How much energy does my family use?

- Does my utility company offer net metering?

- Do I qualify for tax credits?

- Are there any rebates in my area?

The following are examples of when solar energy may not save you money.

- Your roof size will not allow for enough solar panels to offset your energy use.

- Your utility company does not have a net metering program.

- Trees shade your roof.

Mounting Solar Panels

Solar panels are usually roof-mounted, but if you have a shortage of roof space or your roof is not suitable due to obstructions, such as dormers, chimneys, vents, etc., the solar panels can be mounted on a pole or installed on the ground. Ground-mounted systems can be good for homes with large yards.

SUMMARY

While solar can be a beneficial, economical, and sustainable energy source for many, not every home is built for a solar system. It is important to assess all of the factors that determine whether or not solar will be a good fit for your home. You can begin with assessing your own personal energy needs, costs, and your local electrical rates, as well as assessing the location of your house and roof. There may be other options like mounting the panels on poles rather than the rooftop.

Today, there are many different types of solar panels that can accommodate a variety of roof types, expanding to more households the benefits of solar—the topic of the next chapter.

CHAPTER 3

BENEFITS OF GOING SOLAR

I hear and I forget. I see and I remember. I do and I understand.

CONFUCIUS

This chapter will cover the following topics:

- How much will I save on my utility bill?
- Is maintenance expensive?
- Will my home's value improve with a solar system?
- Is there a federal tax benefit?
- Benefit of property tax.
- Money with incentives and rebates.

Utility Bill Savings

Financial returns and lower monthly utility bills are major incentives for going solar. The amount of money you save will depend on the size of your solar system, and your electricity usage. You may also receive solar credits or money for the extra energy that you export back to the utility grid. The savings depends on several of the following factors:

- electricity consumption and local electricity rates

- system size

- purchase or lease of system

- direct hours of sunlight

- size and angle of roof

- solar credits offered

Low Maintenance Cost

Solar energy systems do not require a lot of maintenance, yet it is important to keep the solar panels clean. At least a couple of times per year, simply wash your panels with a hose. However, it is extremely important that you do not wash them when the sun is out or when the panels are hot from the sun. It should be done in the evening or early morning (Chapter 10, "Maintain Your Solar System").

The inverter and cables are usually the only equipment that need to be replaced because they are continuously working to convert solar energy into electricity. It is best to find out from your solar provider how long the inverter and cables will last, and if there is a warranty (Chapter 8, "Your Warranty Agreement").

Increase Your Home Value

A solar energy system may increase a home's value and may

sell the home faster. It all depends on a few things, such as the location of your home, the size of the installation, and the value of your home. For more information, contact a real estate professional.

My Investment Tax Credit

In the United States, one of the great incentives of installing solar panels is the federal tax credit, which is also known as the Investment Tax Credit (ITC). The ITC deducts from your federal taxes a percentage of the cost of installing a solar energy system.

In December 2020, Congress passed an extension of the ITC, which provides a 26% tax credit for systems installed in 2020 through 2022, and 22% for systems installed in 2023. The tax credit expires starting in 2024 unless Congress renews it.

One of the main requirements for the ITC is the solar system must be "placed in service" during the tax year and generate electricity for a home located in the United States. This means that the solar system has to be completely installed, connected to the grid, and the homeowner is able to use the system. The following are the eligibility requirements to qualify:

- You must own the solar system.

- If the solar system is leased or there is a Power Purchase Agreement (Chapter 5, "Power Purchase Agreement"), the tax credit will go to the company that is leasing the system or offering the Power Purchase Agreement.

- To receive the full 26%, the solar system must be "placed in service" before the end of the day on

December 31, 2020, 2021, and 2022.

- To receive the full 22%, the solar system must be "placed in service" before the end of the day on December 31, 2023.

- The year the system is financed or purchased is when the credit can be applied to pay off the taxes paid. For more information, consult with your tax professional.

Not everyone is eligible for the federal tax credit. The following are some reasons for ineligibility:

- You do not pay federal income taxes.

- You are on a fixed income.

- You are retired.

- If you only worked part of the year, you may not pay enough taxes to take full advantage of this credit. For more information, consult with your tax professional.

My Property Tax Credit

The State of California does not provide a sales tax exemption for solar panels, yet it does offer other tax incentives in the form of 100% property tax exemption on solar installations. Installing solar panels may increase the value of the home, which may lead to an increase in property taxes.

In the United States, you can find out if your state will increase your property taxes by going to the Database of State Incentives for Renewables & Efficiency, http://www.dsireusa.org/.

Note:

- *Systems installed before December 31, 2019 were eligible for a 30% tax credit.*

- *This is a guide for going solar and not tax advice or other financial advice.*

World of Incentives & Rebates

Incentive programs pay to produce electricity. In the United States, there are several states and utility companies that offer incentives. For example, the Department of Energy offers as much as 30% in savings through rebates and tax credits. This information can be obtained from the Database of State Incentives for Renewables & Efficiency, https://www.dsireusa.org.

Rebates are offered by many states, cities, and solar companies. Before you sign a contract, ask your solar provider about their rebate options. If they do not have options, inquire about the rebates that are offered by the state and city.

SUMMARY

There are many benefits of going solar—saving on monthly electric bills, increasing the home value, and tax and property credits. All of these positive aspects, along with low maintenance costs, make solar systems an attractive option for homeowners across the nation. Of course, not every homeowner will enjoy the same benefits to the same degree since there are local circumstances and conditions that determine each situation. The next chapter will introduce some of the potential disadvantages of going solar that every homeowner should know.

CHAPTER 4
SOME DISADVANTAGES

Perfection is not attainable, but if we chase perfection we can catch excellence.

VINCE LOMBARDI

The energy production in solar power only takes place when the sun is shining. Solar panels are dependent on sunlight to effectively gather solar energy. When the sun is not shining or when a cloud goes overhead, the efficiency of the solar system drops. This could be considered a disadvantage.

To have a better understanding of all the available options when considering a solar system, here are some examples of some disadvantages.

- The initial cost of purchasing a solar system may be high. To keep costs as low as possible, you can set aside money, ask about possible options, and obtain multiple quotes.

- Severe weather like lightning storms, hurricanes, and hail may damage the solar power equipment. There are some insurance companies that will cover these type of events. You may want to consider consulting with your insurance company and inquire whether the equipment damage will be covered in the event that the weather damages your solar equipment.

- A few cloudy, rainy days can have a noticeable effect on the energy system, causing the efficiency of the solar system to drop.

SUMMARY

Like any technology or innovation, solar systems do have some disadvantages. They are entirely dependent on the sun shining, which does not always happen; the initial cost to purchase and install a system can be high; and damage that requires repairs may occur. These disadvantages, of course, should be weighed against the benefits (Chapter 3, "Benefits of Going Solar") when you make your assessment and final decision. The next chapter will cover financing a solar system in detail—shedding some light on all the options that are available to you.

CHAPTER 5

HOW TO FINANCE SOLAR

I'd put my money on the sun and solar energy. What a source of power! I hope we don't have to wait until oil and coal run out before we tackle that.

THOMAS EDISON 1931

This chapter will cover the following topics:

- Ownership of a solar system.
- Government program called PACE.
- How to lease a solar system.
- What is a Power Purchase Agreement?

Homeowners have multiple options to choose from when deciding to go solar. Some homeowners purchase the entire system by paying with cash or by financing with a loan. There are other options as well, in which homeowners can essentially rent solar panels from a company or even allow power companies to install their own systems and just pay for the power.

Deciding which option is the best will depend on individual circumstances and what choices are available in local areas.

Purchase My Solar System

When you purchase a solar energy system, you pay the cost of the entire system. This can be done in cash or financing the system with a solar loan. Either way, you will pay the cost of the entire system. You will be able to use all the power the system produces, and you may be eligible for federal, state or local tax credits including other incentives.

If you decide to finance the system, the following are some recommended questions:

- Will I have to pay any money upfront?

- What is the annual percentage rate?

- How are the payments calculated?

- Will the payments change during the finance term?

- Is there a balloon payment?

- How long will I have to pay?

- Will a lien be placed on my home?

- How long do I have to cancel the financing?

PACE Program

There is a special type of financing program that your county or state may participate in called Property-Assessed

Clean Energy (PACE), and it is available in areas where local governments have allocated funds. Solar providers or home improvement companies that sell solar systems and other efficiency improvements offer PACE financing. They are paid back through an additional payment on property taxes and are sponsored by your city or state.

The following is the process for PACE:

- A county, local, or municipal government passes legislation that establishes a PACE program and makes funds available to investors.

- An authorized PACE lender provides those funds to homeowners.

- Homeowners repay the finance company through an assessment attached to their annual property tax bill.

Here are some recommended questions:

- What is the difference between the interest rate of PACE versus a traditional loan?

- Can a PACE lien affect my ability to refinance or sell my home?

- Are there any fees for early payoff?

Lease the Solar System

A solar lease is a financing option that allows you to have solar panels installed without having to worry about the upfront costs. The term of a solar lease is usually 15 to 20 years. The solar company owns the panels during the term and is usually responsible for maintenance. They also provide a 20-year warranty. This is known as a third-party solar lease because the panels are owned by a third party and not the homeowner. The guarantee is that you will never overpay for electricity, and if it does not generate the amount of electricity that was estimated, you will receive the difference to compensate your bill.

The following is the process to lease a solar system:

- You sign a contract to use a system the company installs on your home.

- You use all the power the system produces.

- You pay a monthly leasing fee that may increase over time.

- You are not eligible for tax benefits or incentives.

- You pay the company a certain rate for electricity.

- You do not have to maintain the solar panels.

- You do not own the system—when the lease ends, the system will be taken away or you will have to purchase the system.

Power Purchase Agreement

A Power Purchase Agreement (PPA) is similar to a solar lease, but instead of a leasing payment, you agree to pay for the power the company's panels produce. In the United States, PPA's are currently available in California, Connecticut, Massachusetts, Maryland, Nevada, New Jersey, New Mexico, New York, Pennsylvania, and Rhode Island.

The following is the process for a PPA:

- You sign a contract to buy power from a system that a company installs on your home.

- You do not get all the power the system produces.

- You pay for power at a rate the PPA provider sets, which may be lower than the utility company's rate.

- You are not eligible for tax benefits or incentives.

- You do not have to maintain the solar panels.

SUMMARY

Perhaps the most confusing, unfamiliar aspect about solar systems for homeowners is financing. There are several different options that are available, including outright purchasing a system, taking advantage of the PACE program, signing a Power Purchase Agreement (PPA), or leasing.

The decision of how to finance is an important one, so be sure to take your time, ask questions, and do your homework. Buying, leasing, or signing any agreement is a long-term commitment. For many homeowners, selling energy is part of their financial calculation. The next chapter will provide information about how to sell energy and explain the important terms and factors.

CHAPTER 6

SELL SOLAR ENERGY

The purpose of information is not knowledge. It is being able to take the right action.

PETER DRUCKER

This chapter will cover the following topics:

- Special type of billing arrangement.
- How will my utility company bill me?
- How is my energy recorded?
- Benefits of net energy metering.
- What is a true-up statement?

There are several important steps along the path to installing solar. One step is connecting your solar system to the utility electric grid. The amount from selling solar energy back to the grid is based on the local market value for electricity.

A utility company is responsible for connecting a solar system to the electric grid, and depending on the utility company, it will provide solar credits for the exported solar energy. This process is called solar interconnection (Chapter 9, "Interconnection Agreement").

The energy you produce with your solar system, minus the energy you consume at home equals net energy. When your system generates electricity, that electricity flows into your home and is consumed right away. When your solar panels produce more electricity than your home needs, the excess electricity is sent out to the local grid.

Not all utility companies have net metering incentives, some offer no credits, and some charge solar customers additional fees for access to the electric grid. For further information, contact your local utility company and solar provider.

Net Energy Metering

The Net Metering Agreement is a contract that you enter with your local utility company when your solar system is connected to the grid. The practice of selling power to your utility company is known as net metering. It is a way of using the grid to store the energy produced by your solar system for later use. It is like having the grid serve as a humongous battery. The "net" part of the term means that the homeowner pays the "net" amount for the electricity used by the house, minus the extra electricity sold back to the grid. Simply, net metering subtracts solar production from electricity consumption, and you are billed the difference.

Net Energy Metering (NEM) is a special billing arrangement with your utility company that allows you to receive a credit on your electric bill when your solar system sends electricity back to the grid.

With NEM, your electric meter keeps track of how much electricity is consumed by your home, and how much excess electricity is generated by the system and sent back into the electric utility grid.

The way it works is when solar panels produce more electricity than you are using during the day, the electricity is sent back to the grid running your electric meter in reverse. However, when your home uses more electricity than your solar panels are producing either at night or on cloudy days, you pull electricity back from the grid running your meter forward. If there is any excess energy sent to the grid, you may be compensated in the form of credits on your electric bill. At the end of the month or year, you are billed the net of what you put onto the grid and what you took off the grid. You use what energy you need, and you have a market ready to buy your surplus energy, which lowers your overall electric bill.

Even though net metering sounds great, not every state in the United States allows net metering. Among those that do, the policies vary—some are more favorable for the consumer than others.

There are 44 states, as well as the District of Columbia, that have mandatory net metering rules. Alabama, Mississippi, Tennessee, and South Dakota do not have net metering policies. Texas and Idaho do not have mandator state-wide net metering rules, but some cities and energy providers in those states offer net metering.

In addition, it is important to understand the fine print because not all net metering programs are the same. Be aware that policies can change.

The following are some questions to ask your utility company and solar provider:

- Is net metering credits rolled over from month to month?

- What happens to unused credits at the end of the year? Do they expire? Are they paid out at a lower rate than monthly surpluses?

- Does your utility company charge a monthly connection fee?

- How much will a utility company pay for the excess power you produce through net metering?

- Does your utility company or state impose limits on how much surplus electricity you can sell back to the grid?

Many states pay retail rates for surplus electricity—so the utility pays you the same rate it charges when it sells its electricity to you. However, about 10 states pay consumers "avoid costs rates," which are basically what it cost the power company to produce its electricity—and these are lower than the retail rate.

— Billing

After your solar system is connected to the grid, your monthly electric bill will breakdown how much energy you received or consumed from the utility, and how much your solar system sent to the grid. In general, most homes will produce excess electricity in the summer months and will use more electricity from the grid in the winter. When your solar power system generates more electricity than you use over the course of a month, you will receive credit based on the net number of kilowatt-hours you gave back to the grid. If you produce less electricity then you use in a month, you must buy electricity from your utility to make up the difference. In these instances, you would pay for the electricity you use minus any excess electricity your solar panels generated.

Over a 12-month period, you will pay for the net amount of electricity used from the utility grid that was above the amount of electricity generated by your solar system. In other words, if you use more energy than you produce, you will have to pay for some electricity. On the other hand, if you use less energy than you produce, you will not have to pay for electricity.

You have the option to pay the utility company for your net consumption monthly or settling your account every 12 months. Contact your utility company for billing options.

— Bidirectional Meter

A bidirectional meter measures the flow of electricity in 2 directions. It measures how much solar energy it has fed back into the grid by spinning backwards. If you generate more electrical energy that you used from the utility's electrical system, then you will receive solar credit if your utility company is part of the net energy metering program. Many residential solar systems that maintain a connection with their utility company require a bidirectional meter.

— Metering Aggregation

Metering aggregation allows a single customer with multiple meters on the same property or an adjacent property to receive the benefits of net energy metering. Contact your utility company for their requirements.

— Benefits of Net Energy Metering

- Allows customers to zero-out their bills.

- May credit customer's accounts at full retail rates.

- Accurately captures energy generated and consumed.

- Customers can generate their own electricity cleanly and efficiently.

True-up Statement

A true-up statement is an annual bill. It is the net electricity usage for the year and summarizes electricity charges and credits for each month into the annual bill. You will see a breakdown of how much energy the company has credited your account, and how much you have consumed. You will pay the bill annually, although you may receive a quarterly or monthly statement that breaks down your usage and credits.

If the solar system exports more energy in a month than the home draws from the grid, the utility company carries over any unused credit at retail rates to the next month. Any remaining credit at the end of the yearly billing cycle, where you have exported more to the grid than what you have used from the grid, will be paid out at a lower rate. This is if the system does not cover what the home draws from the utility grid. For more information, contact your utility company.

SUMMARY

For any homeowner who is connected to the grid, going solar does not end your relationship with the utility company. You will still receive bills or statements, and it is important to understand how metering works.

The ability to produce excess energy depends on your

system, where you live, and your energy consumption patterns. Knowing how energy is measured and getting a realistic picture of how much energy you will produce and sell is important for homeowners considering solar. Local utility companies can provide more specific information that is useful. Finding reliable information is essential when considering purchasing a solar system. In the next chapter, you will learn about scammers in the solar industry.

CHAPTER 7
SCAMS DO EXIST

Beware of he who would deny you access to information, for in his heart he dreams himself your master.

SID MEIER

This chapter will cover the following topics:

- People who give the wrong information.
- What is the process of signing electronically?
- How do I protect myself from scams?

The scamming industry makes millions knocking consumers down in a domino-like effect. There are always a couple of bad apples, and they exist in the solar industry.

Solar Providers Who Mislead

There are many people who have no knowledge about what they can expect from solar salespeople, and for this reason, it is very easy for conniving people to take advantage of these unsuspecting individuals. There are stories from homeowners, who were promised gifts once they signed, such as paid vacations, household equipment, or money. Yet, there was nothing written on paper to confirm the verbal agreement. These trusting homeowners were left standing in the dust holding on to empty promise. There are no more calls, emails, or texts from these scam artists when prior to the homeowner's signature, they heard from them 2 to 3 times a day.

There are hundreds of solar scam ads on the internet. Social media sites have become common ground for promoting ads of special low-cost or zero upfront-cost solar programs. Do not let them fool you. There is no special government or municipal programs that work this way. It does not exist. These ads are used by private companies to generate leads—or consumers— who they think are likely to buy their products or services.

In this scam, a consumer fills out their information online through a sign up form to see if they qualify. Then, a private company often follows up with some type of solar loan or financing product.

Here are some examples of what to look out for in solar scam ads on the internet.

- An exaggeration of the upfront costs of a solar installation making it look like purchasing a solar system is the only option.

- Ads may disguise the loan product as a solar installation or other home improvement, such as a roof replacement. Sometimes, it states that it is a special program sponsored by a government or utility company.

- A claim about a program funding available only in your area and for a limited time.

- Solar ads on Facebook advertising 100% free panels and fake stimulus programs.

- Ads that urge you to share your address, contact information, or utility bill to "see if you qualify." This information is used to sell you a product or service.

The salespeople from unscrupulous companies are usually the main point of contact for homeowners. They are the door-to-door solar salespeople. Be very cautious. The following are some red flags.

- Claims that a special financing rate or incentive program will only be available for a short time. This is to pressure you into signing a contract on the spot.

- Engages you in a conversation and attempts to sell you items you are not ready to buy.

- Claims that they work for the electric company, and a review of your energy bills is required.

- Requests an inspection of your roof for a roof survey.

- Claims that solar energy is free and energy costs will be lowered.

- Claims that a single brand of technology is the only option available.

- Claims that money will be made with the installation of solar panels.

- Requests that you sign on an electronic device, such as a cell phone, tablet, or iPad to gain access to your property.

Note:
Do allow anyone to sell you a solar system on the phone. You do not know who is on the other line.

Electronic Signing

As the world is becoming paperless, more contracts are being signed electronically—which is sometimes called e-signing or e-signature. The problem is that even though there is no paperwork, the fine print lurks in the background, and it may come back to haunt you.

For example, if you signed on your cell phone, tablet, or iPad, the salesperson may not have mentioned that the document contained a multiple-page contract with fees, and now your e-signature provided access to your credit report.

I spoke with many homeowners who stated that the signature on the document was not their signature. Some remembered receiving the document via email but would say that it was not their signature. The primary complaint was that the electronic signing process was not explained properly.

It is extremely important you have your own personal email before you sign any documents. The email ensures you receive your finance contract, solar agreement, and any other documents pertaining to your solar system. If you do not have an email, ask someone you trust to explain what is the purpose of an email. You can create an email for yourself or have the the person you trust create one for you. If the email is created for you, make sure you write it down, including the password, and save it. Then, as soon as possible, open your new email and change the password to secure your information.

When you have your own email, and you want to use it to receive your solar documents, do not allow anyone to tell you that you cannot use your own email. It is up to you and nobody else to make that decision for you.

Note:

- *If you do not understand the information, always ask to have it explained until you understand. Do not allow anyone or anything to stop you from asking any questions.*

- *If you have any doubt, stop the process. Inform the salesperson that you need time to think about it, and you will contact them at a later time.*

- *You have the option to request a paper contract and sign with a pen. It is called a "wet signature."*

How to Protect Yourself

There are many people who want to know if they can get their money back after they signed a contract. The main problem is that when the innocent-looking box was checked that says, "check if you agree," they agreed to give up their consumer rights. Most likely, the contract contained an arbitration clause, which means that some or all of their rights were waived (Chapter 8, "Arbitration"). Some of these rights include:

- The constitutional right to a jury trial.

- The right to appeal.

- The right to a judge.

- The right to bring a class action.

When you sign your rights away and "agree" to have your case decided individually in an arbitration proceeding, you may not have any legal options. Here are some tips to protect yourself.

- Take your time and do not feel pressured.

- If you feel rushed, stop the salesperson even if it is in the middle of a sentence. It may be a solar scam.

- Inform the salesperson that you will think about their offer.

- Ask for their business card, and call the number.

- Be cautious about big promises of energy savings, referrals, rebates, or tax credits.

- Never give out personal information to a stranger like bank account numbers, birth dates, or social security numbers.

- Monitor your credit. It is illegal for anyone other than you to make a hard inquiry on your credit report without your permission. The 3 credit bureaus are TransUnion, Experian, and Equifax.

- Research stories about solar scams in your area.

- Know all the terms before signing a contract.

- Insist the arbitration clause be removed from the contract.

- Research solar companies and request pricing estimates.

- Find out how much power the solar system will generate by going to PVWatts Calculator by the National Renewable Energy Laboratory, https://pvwatts.nrel.gov/.

- Visit websites such as the Federal Trade Commission (FTC), https://www.ftc.gov, and the Consumer Financial Protection Bureau, https://www.consumerfinance.gov.

SUMMARY

It is unfortunate, but solar scammers do exist. They will make misleading promises, provide false information, and pressure customers to sign before they understand all of the consequences. The best defense against a solar scam is good information. So, do your homework, ask questions, and wait to sign until you feel confident. The next chapter will cover contracts, warranties, and solar providers—all important topics for anyone considering solar energy to know about before signing the dotted line.

CHAPTER 8

PROVIDERS, CONTRACTS & WARRANTIES

Our lives begin to end the day we become silent about things that matter.

MARTIN LUTHER KING JR.

This chapter will cover the following topics:

- How to choose your solar provider.
- The importance of the solar contract.
- Understanding the types of warranties.

Choosing Your Solar Provider

The solar providers are a team of accredited specialists. They handle all aspects of your solar power project—from designing the system to suit your needs through the installation and on-going services. A solar provider usually submits all the required paperwork to your utility company on your behalf. Sometimes, they provide financing directly in the contract.

Finding a qualified solar provider is key to getting a productive solar system at an affordable price.

There are many types of companies involved in a solar project that work behind the scenes, such as solar equipment manufacturers, solar installers, and the finance companies. Some providers have in-house solar professionals and others are subcontracted. Take your time and compare your options before making this long-term investment. Here are some tips:

- Research for reputable companies by checking their company reviews. Search for satisfied customers in places like the Better Business Bureau, https://www.bbb.org and EnergySage, https://www.energysage.com.

- Make sure they are licensed and bonded.

- Ask for at least three references and call them.

- Obtain at least three quotes with specifics about the solar system, such as the full cost of installation, whether the system is guaranteed to produce a certain amount of energy, and what warranties apply to the equipment.

You can check a company's history with your state and local consumer protection agencies and the state contractors licensing board. For example, in California, there is the Contractors State License Board, https://www.cslb.ca.gov.

The following questions are recommended to ask the solar provider:

- How long have you been in business?

- What is your experience?

- How many solar systems have you installed?

- Can you provide customer references?

- Do you operate in multiple states?

- What are your licenses and certifications?

- Are your installers in-house or third party?

- Do you handle the paperwork for incentives?

- Do you offer maintenance service?

- Are you a member of any solar organizations?

- What warranties do you offer?

Your Solar Contract

The solar contract is a document between you and your solar provider. It is commonly called a home improvement contract. Before you sign, read it thoroughly. Make sure everything you were promised is in the contract, and additional services should be separate line items in the contract price. Be aware that after you sign the contract, you have a three-day right to cancel (more information on the following pages titled, "Three-Day Right to Cancel").

You and the contractor must avoid misunderstandings. Asking questions upfront is crucial so do not hesitate. If there is any information not clearly defined in the contract, ask the solar provider to clarify each point. Here are some questions to ask before entering into any agreement.

- What is the total cost of the solar system?

- What is the system size?

- How much is the total cost if I add storage?

- What is the timeline for this investment?

- How much do I pay upfront?

- How long will I be paying?

- Will a lien be placed on my home?

- How much electricity will it generate yearly?

- Is there a minimum guarantee for production?

- Are there any other guarantees?

- Will my system have net energy metering?

- If net energy metering is not available, how will I be compensated for excess electricity?

- If there is a blackout. what will happen to my system?

- Can I get out of my contract?

- How much will it cost to get out of my contract?

- What happens if I want to sell my house?

— Three-Day Right to Cancel

Several federal laws (known as "cooling-off rules") allow you to cancel your solar contract after you sign. You have up to 3 days to cancel your contract. You may cancel the contract by email, post office mail, fax, or deliver a written notice to the solar provider's place of business. The cancellation must be received by midnight of the third business day after you received a signed and dated copy of the contract. Be sure to include your name, address, and the date you received the signed copy of the contract. If your solar provider refuses to cancel the contract, contact your state's local licensing board.

— Senior Citizens Five-Day Right to Cancel

In the State of California, senior citizens (65 or older) have up to 5 days to cancel their contract. You may cancel the contract by email, post office mail, fax, or deliver a written notice to the solar provider's place of business. The cancellation must be received by midnight of the fifth business day after you received a signed and dated copy of the contract. Be sure to include your name, address, and the date you received the signed copy of the contract. If your solar provider refuses to cancel the contract, contact your state's local licensing board.

— Mechanics Lien Warning

A Mechanics Lien Warning is found in solar or home improvement contracts. It states that anyone who helps improve your property and is not paid, may record what is called a Mechanics Lien Warning on your property. It is a "hold" against your property filed by an unpaid contractor, subcontractor, laborer, or material supplier and is recorded with the county recorder's office.

Be aware that even if you pay your main contractor in full, unpaid subcontractors, suppliers, and laborers who helped to improve your property may record a Mechanics Lien and sue you in court. If a court finds the lien is valid, you could be forced to pay double the same job or have a court officer sell your home to pay the lien.

Prior to the Mechanics Lien Warning, each subcontractor and material supplier must provide you with a document called a "Preliminary Notice" to continue their right to record a lien. However, this type of notice is *not a lien*. The purpose of the notice is to let you know that the person who sends you the notice has the right to record a lien on your property if he or she is not paid. This notice can be sent up to 20 days after the subcontractor starts work or the supplier provides material. This may be a problem if you pay your contractor before you have received a preliminary notice.

You will not receive Preliminary Notices from your main contractor or from laborers who work on your project. The reason is because the law assumes that you already know they are improving your property. Here are some tips on how to protect yourself from a preliminary notice.

- Obtain a list from your solar provider of all the subcontractors and material suppliers.

- Ask your solar provider when the subcontractors started work and when the suppliers delivered materials.

- Wait 20 days and pay attention to the preliminary notice you receive.

- For cash deals, when the solar provider says it is time to pay for the work of a subcontractor or supplier, write a joint check payable to both the contractor and subcontractor or supplier.

— Arbitration

Arbitration is a procedure in which a dispute is submitted to one or more arbitrators who make a decision on the dispute. In choosing arbitration, the parties choose a private dispute resolution process instead of going to court. In a solar or home improvement contract, there is a document called "Arbitration of Disputes." If you initial this document, you are giving up any rights to have the dispute fought in a court or jury trial. You give up the right to discovery and appeal unless those rights are specifically included in the "Arbitration of Disputes." If the contractor and homeowner do not each initial the "Arbitration of Disputes," then it is agreed that neither party agree to arbitrate and the "Arbitration of Disputes" shall not be part of the contract.

Your Warranty Agreement

A warranty agreement is a promise that something will be done a certain way. Warranties protect your solar panels and other related equipment. If anything happens to your solar panels, having a warranty helps you hold the manufacturer and solar provider accountable without any additional cost to you. It is very important you understand what is covered by the solar and manufacturer warranties. That way, if a problem arises, you will know exactly what to expect.

The following information is on 3 types of warranties: product, power production, and installation labor.

— Product Warranty

The product warranty guarantees the physical integrity of your solar panels and inverters. They are offered through the manufacturer, not your solar provider. For example, if a connection on one of your panels fail, and it affects its electricity production, that panel would be replaced under the product warranty. Most manufacturers offer a product warranty between 10 to 25 years. Inverters come with product warranties that range from 10 to 15 years. String inverter manufacturers offer product warranties from 10 to 15 years, while micro inverter manufacturers offer warranties for 25 years.

— Power Production Warranty

The power production warranty is offered by the panel manufacturer, and it guarantees that the solar panel will produce at 80% of its nameplate capacity for a minimum of 25 years. This means that the capacity is the maximum output a generator can produce from a power plant or generating facility. For example, if a power plant has a nameplate capacity of 500 megawatts, the plant is capable of producing 500 megawatts operating at continuous full power. Just like the product warranty, the power production warranty is offered by the panel manufacturer. While details vary from manufacturer to manufacturer, the power production warranty assures you that your system will continue to produce power over its lifetime.

— Installation Labor Warranty

The installation labor warranty covers the installer's workmanship, including their electrical wiring work and any roof penetration they make to attach your solar array to the roof. This warranty is offered by your solar provider, and it typically ranges from 3 to 10 years. Some providers will offer an option to upgrade to a longer labor warranty while others cover labor costs for any equipment servicing required during their warranty period. It is important to read the workmanship warranty thoroughly before signing a contract with a solar provider. Make sure to ask for a warranty certificate in case there is any kind of malfunction or issue with the solar panels.

SUMMARY

There can be a world of difference between solar providers, contracts, and warranties that are offered. It is up to you to gather as much information as you can to ensure that you choose the best options available. Getting multiple quotes from different providers, verifying their credibility from local licensing boards, and asking lots of questions about the contract and warranties is the best way to make sure you get the right product at a fair price.

If the warranty agreement and contract seem overwhelming, revisit this chapter and identify any terms that are not spelled out or seem different from the standard elements presented here. The next chapter will explain what the installation process will look like, which is helpful for any homeowner who is ready to move forward with a solar provider.

CHAPTER 9

BEHIND THE PANELS

Genius is one percent inspiration and ninety-nine percent perspiration.

THOMAS EDISON

This chapter will cover the following topics:

- Home and roof inspection.
- Design process of the solar system.
- Importance of the permitting process.
- Who orders the solar system equipment?
- The job of each solar component.
- Inspection and interconnection procedures.
- When will my solar credits begin?

Every home is different, which means a solar panel installation is always a custom job. Prior to an installer placing the panels onto your home, they first need to assess your property.

Site Survey

The first phase of any residential solar installation is the site survey. Once you sign your solar contract, an installer will visit your home to make sure the solar design is customized for your unique energy needs. The installer could be an in-house employee or an independent contractor working with your solar provider.

The roof of your house is the most important. The installer will take the roof's measurements, evaluate its structural integrity, and collect readings on shading and sunlight availability. In addition, the roof's structure will be evaluated to make sure it is able to support the additional weight of the solar panels, including your home's electrical panel to determine if it will be able to connect to your electrical system. If the electrical panel is older or if you have recently completed a home improvement project that increased the number of electrical connections in your home, an upgraded electric panel or main panel upgrade (MPU) may be required to accommodate the inflow of solar energy. For more information, contact your solar provider.

The Design

After the site survey, your solar provider will recommend a design for your home. The design focuses on each solar array. The panels are organized into arrays on rooftops or on ground mounts to ensure the panels are organized into a pattern that allows them to be at their highest efficiencies and production levels, including all electrical and permitting codes are followed.

Ultimately, you have the final say over the design of your home, so do not hesitate to ask questions and work with the installer on any design or performance inquiries you may have.

Your solar provider will request your approval before the design is finalized and applies for the necessary permits. Once it is approved, the installation process will move on to the next phase.

Permits

Once the design is complete, the solar installer visits the appropriate government offices (i.e., planning and zoning commission) to petition for necessary construction permits before the installer can go ahead with the installation. Installers make the process look simple and quick; however, it is considered a major construction project and requires special permitting from the city, county, or other jurisdiction offices.

You need the proper permits and permission from your municipality or you will risk fines. Your solar provider will handle the entire permitting process for you, but it is important to make sure you obtain copies of the permits once available and keep them for your records. You may need copies when selling your home to document that all necessary approvals were received, and copies of the permit(s) can be a requirement for certain incentive and financing programs.

Building permit regulations are specific to residential areas, so your installer will be an excellent source of information about the different requirements. For instance, some states allow homeowners to install solar panels across their entire roof, while others require 3 feet of clear space surrounding the solar panels for safety access. Obtaining the right permits can take weeks, and it may take a little bit longer if you run into any issues.

Interconnection Application

The interconnection agreement is a written notice to a utility company of plans to construct, install, and operate a solar system that will be connected to their grid. After the utility company receives the required documentation, the application is reviewed for approval. Once the initial permits and interconnection agreement are in hand, the installation of the solar panels, inverter, racking system, and wiring can begin.

The interconnection agreement states that you will be charged for your net power usage. Net metering (Chapter 6, "Net Energy Metering") uses a bi-directional meter that spins forward when energy is taken from the grid. When excess energy is produced, the meter spins backward. When your solar panels generate more electricity than you are using, the excess kilowatt-hours (kWh) are exported to the grid, and the meter will spin in reverse. As a solar homeowner, this allows you to export (sell) excess energy to the electric grid when your system produces more than you use, and you can purchase electricity from the grid when you need it.

In the interconnection process, there are 2 parts: applying for interconnection and receiving permission to operate (PTO) your solar panels. Utility companies will not allow just any solar energy system to connect to their grid. They need to ensure that your solar energy system meets necessary electrical safety standards and their net metering guidelines.

Interconnection applications often require information about your property, your electricity usage history, and the specifics of the system you are looking to design. Applications for interconnection may be submitted by the utility account holder, but most installation companies will submit it on your behalf. If there are any red flags or missing information in the application, a utility may deny interconnection to the grid and request updates or resubmission.

Once your electric utility grants approval for the installation, you and your installer can move forward with the installation process. After your solar system is installed on your property and your local government has finished their own inspection process, the final step towards connecting to the grid is PTO.

As the first step towards PTO, a utility representative will often be sent out to your property to examine the system. During this visit, the rep usually looks at the inverter, the connection at the electrical panel, and the capability of the system. The rep usually installs an additional meter or upgrades an existing one. This way they can track your solar electricity exports to the grid, therefore, enabling you to take advantage of the utility's net metering incentive.

Following the inspection and meter upgrade, you will receive official PTO documentation notifying you that you can officially turn on your solar system for electricity production. At this point, if you are a customer of a utility that offers net metering, you will receive credits for this electricity that you can apply towards a future electric bill. Make sure you talk to your solar provider about how long the interconnection process is expected to take.

The Equipment

Once the paperwork and permits are approved, your solar provider will place an equipment order through their primary distributor. This will include 2 major components: inverter(s) and solar panels. The equipment order will be based on the approved design which will be finalized after your site survey. Once the equipment has been received, your home is ready for your solar panels to be installed.

The Installation

Once the installer has gained approval to proceed with the installation and depending on the size of the system and other factors, your specific installation date will depend on your solar provider.

The installers often start by ensuring the tiles or shingles on the roof are attached appropriately, and they will set up mounting equipment. They will connect the electrical wiring that will connect to your general power system and electrical panel that was inspected during your site survey. Once the electrical wiring is done, racking will be installed to support your panels, and then the panels will be placed on the racking.

Lastly, the inverter system will be connected to the panels to convert the direct current (DC) energy coming out of the panels into alternating current (AC) energy that can be used by your home and the electric grid.

The Inspection

Before you can turn on your panels, an inspector from the local building and electrical department will examine your home. Your solar provider will send an installer to meet the inspector, and show all of the technical details they will need to be reviewed, including all the interconnection points and electrical systems. If everything was installed correctly and meets local regulations, the inspector will give written confirmation that the inspection has passed. Once it has passed, your solar system is officially ready to be connected to the grid. Ask your solar provider how long the inspection will take to complete.

Turn On Your Solar System

After the final inspection has passed, it may take weeks for everything to go through as you wait to hear back from the utility company. Your solar provider works with the utility company to request the official "Permission to Operate" (PTO) documentation. The PTO document is issued by the utility company for the solar system to generate power. This document is usually emailed to you and/or your solar provider, although there are a few utility companies that have mailed this document.

PTO means that the panels are in full operation working in conjunction with the utility grid and feeding into the energy meter attached to your home. You are now generating clean, renewable energy, and you will begin receiving your solar credits if your utility company is part of the net metering program (Chapter 6, "Net Energy Metering").

SUMMARY

Once you have selected a provider and feel comfortable signing the contract, the next step is installation. Because installing a solar system involves multiple factors—the provider, the installer, the utility company, and potentially others—it can be an involved process. Your roof or property will need to be assessed in order to design a plan, permits will need to be secured, and then the system will need to be connected to the grid. The homeowner is not necessarily deeply involved in the installation, approval, and connection process; however, it is good to know what to expect. The next chapter will cover any additional aspects that homeowners may want to know.

CHAPTER 10

OTHER SOLAR TOPICS

The decision is your own voice, an opinion is the echo of someone else's voice.

AMIT KALANTRI, WEALTH OF WORDS

This chapter will cover the following topics:

- My energy production.
- Purpose of a battery backup.
- How do I take care of my solar system?
- Can my solar system be off-grid?
- How long do solar panels last?
- The writer's final thoughts.

You never know when there will be changes in the solar industry. It is best to keep yourself informed. For example, in 2020, Congress passed a two-year delay of the Investment Tax Credit (Chapter 3, "My Investment Tax Credit"), which means that the tax credit was extended 2 additional years.

View Your Production

Solar energy monitoring is the process of connecting your inverter to the internet. It collects and organizes large amounts of performance data, making it easy to monitor your system. The data is typically available via an app on a smartphone, tablet, or computer.

There are 2 ways to monitor your system. The first option is with your internet or Wi-Fi. You will need to maintain a Wi-Fi connection to manage your energy use from your devices like your smartphone, however, Wi-Fi connections are unreliable. Every now and then, you will have to replace your router or change to a different service provider, and if you do, your inverter will have to be reconfigured to reestablish communication with your Wi-Fi.

The second option is a cellular modem card or cell-based connection (cell card). This provides networking without relying on an internet connection in your home. It will ensure that all your data is being captured even when your Wi-Fi fails. You will experience an uninterrupted access to view your energy consumption via an app or website. The cell card allows you to keep your internet use separate from your entertainment and/or business.

Battery Storage

A battery storage is a device that reserves energy for later consumption and is charged by a connected solar system. Because the solar panels can only produce power when the sun is shining, the battery storage is used to store unused energy throughout the day for use at a later time. The stored electricity is consumed after sundown during energy demand peaks, or during a power outage.

Storing electricity in a battery bank can serve many

purposes. In most parts of the country, battery storage for residential home owners is mainly used to provide backup power during power outages. When the utility grid goes down and you lose electricity service, you can use a battery system to power some or all your household electricity needs called "loads." The battery backup system works by isolating certain loads from the main utility system with an "automatic transfer switch." You are then able to power those loads with electricity stored in the battery bank.

Loads can range from small (toaster, hair dryer) to large (refrigerator). When the utility grid power returns, the backed up loads in your home then automatically reconnect to the grid. The result is that these "critical" loads receive power even when the grid is down, switching easily between utility electricity and stored electricity from your battery.

Note:
Home backup batteries run on electricity and can be charged by the grid or solar panels. Not all home battery systems can be recharged during power outages so make sure that your solar provider knows this feature is crucial to you.

— Battery Types

There are 2 types of batteries available, and they have different chemistries. Lead acid batteries are generally installed indoors. Lithium-ion batteries can be installed outdoors and have a much wider preferred temperature operating range.

— AC Coupled & DC Coupled Systems

When it comes to the way your solar panels, batteries, and inverters are all wired together, there are 2 main options: AC coupled (alternating current) and DC coupled (direct current).

AC coupled systems require 2 inverters: A common grid-tied solar inverter and a battery based inverter. One inverter converts the solar electricity from its natural DC form to AC. This allows it to flow directly into your home. The other inverter allows the battery to convert the solar electricity back to DC so that it can be used to charge the battery. Typically, the energy produced from your solar system enters your house, and excess electricity not used is stored in the battery.

DC coupled systems are most common when installing both solar and battery storage at the same time. In this setup, the solar electricity is fed directly into the battery system without the need for any changes. The battery and solar system often share one inverter in DC coupled systems, and have an easier time ensuring that your battery is only charged by solar electricity, as opposed to electricity from the utility grid.

When adding batteries to an existing solar system, many systems will be AC coupled. This is done to allow you to keep your existing solar inverter and wiring. A second inverter will be added in addition to the existing solar inverter. It is important to note that AC coupled systems make it more difficult to guarantee your battery is exclusively charged by solar. This is an important eligibility factor for the federal tax credit. For more information, contact your solar provider and tax professional.

Powering your entire home with a battery system can be expensive. This is why many homeowners install a smaller battery bank to power select "critical loads," such as medical equipment or a refrigerator in their home during the event of a grid outage. There are other technical considerations, such as batteries take up room in your home, may require maintenance, and will likely need to be replaced at least once during the lifetime of your solar system.

— Benefits of Battery Storage with Solar

- A reduction in peak demand charges.

- The option to cut down your electric bills.

- The battery will charge during off peak hours and distribute the saved energy during peak times.

- A reliable backup power in case of a major outage.

Depending on the type of battery, it could be placed either inside or outside the house. If placed outside, and depending on the battery chemistry, the battery may need to be placed in a shaded area. Your installer may need to adjust the size of your battery system to accommodate your available space.

Since batteries can be expensive, most people size their systems to only power "critical loads" while the utility service is out. Consult with your solar provider to help you decide which loads you want to power with your battery.

Without a battery, your solar system will not provide electricity to your home during a power outage. This is because solar systems are required to automatically shut off if the grid goes down. This is done to ensure that they do not "back feed" power onto the lines and injure workers who are repairing the electric lines.

Note:
Not all solar installers have experience installing batteries. Ask your solar provider if they have experienced installers for battery installation and make sure that it is put in writing.

Maintain Your Solar System

Solar panels require a little maintenance. They need to be inspected at least once every 6 months for any dirt or debris that may collect. In most cases, solar panels do not need to be washed because rain and snow naturally clean them. In areas with less rain and lots of dust or pollutants in the air, occasional cleaning may improve performance. To make sure your solar panels remain in good operating order, it is recommended to ask your solar provider about their maintenance procedures.

— Maintenance Tips

- Check the solar panels are not being shaded by vegetation or building structures between 9 a.m. and 3 p.m. or your peak production hours. Trim vegetation if necessary.

- Visually inspect the solar panels from the ground for damage. Solar panels have a protective glass front that may break from hail or other causes. If you discover broken glass, turn your system off and call your solar provider immediately.

- Check the solar glass surface for debris, dirt, or bird droppings. Seasonal rain should wash away normal soiling. If you choose to clean the panel surfaces, first verify that there are no broken solar panels in your array. Then, remain on the ground and spray the glass with water from a hose (important to read next page, "Safety Hazards").

- If you live in a desert or highly dusty environment, periodic cleaning is recommended. The same is true if you are near the ocean and receive marine mist that could leave salt deposits.

— Safety Hazards

- DO NOT clean during the middle of the day when the glass is hot. The thermal shock of cold water on hot, tempered glass could shatter the glass. Clean only at dawn or dusk when the panel's glass is cool.

- DO NOT clean or touch broken panels. Solar panels have a protective glass front. Broken panel glass is an electrical safety hazard (electric shock and fire). These panels cannot be repaired and must be replaced immediately. If you have a broken panel, turn your system off and call for service immediately.

- DO NOT clean the solar panels if your inverter reads "Ground Fault Error." Contact your solar provider immediately.

Off-Grid Solar Systems

Off-grid solar means that all your energy needs are from the power of the sun with no help from the utility grid. You are completely reliant on the sun and energy stored in batteries to power your home. In addition, if there is no generator for an off-grid system, electricity will only be available when the sun is shining, and the solar system is producing electricity. Also, the electricity previously generated by the solar system is from a solar storage device—like batteries.

If you do not have batteries to store your energy, you will have less or no electricity when it is cloudy, and you will not have electricity in the evening. With an off-grid system, you will not have access to extra electricity if you need it.

Batteries are very useful for off-grid systems because they can be charged during the day, and you can use the energy at night. This is a good solution for using solar energy all day, but it can be expensive. For more information, contact a solar provider.

Life Expectancy

Solar panels last about 25 to 30 years, however, the installation and the type of material used may affect the lifespan. Here are 3 tips to ensure your solar panels last for a longer period.

Tip #1

Consult with your solar provider about routine maintenance checks. These checks can alert you to any quality degradation in the panels, any issues with the racking attached to the roof, and whether or not the inverter is hooked up properly.

Tip #2

Keep your panels clear of debris and other damaging materials. When your panels are free from leaves, dirt, pollen, dust, bird droppings, and falling branches, this will help them generate the highest amount of solar energy possible. Soiling affects the lifespan of a solar panel. Tiny dirt particles start to accumulate over the surface of a solar panel and may lose its efficiency due to the loss of solar radiation.

Tip #3

Birds may pose a threat because they prefer the dry area below the panel for making their nests and leaving their droppings. Bird proofing is an essential maintenance step and keeping them clean will keep your photovoltaic array in good working order. Here are a few ideas on how to prevent birds from moving on your property.

- Bird Mesh | Designed to seal the area under your solar panels. It clips directly to the panels and runs around the edges of the entire array.

- Roof Spikes | They may not be attractive, but it makes it uncomfortable for the birds to be around your solar panels.

- Plastic Predators | There are fake owls with a head that swivels in the breeze. There are also high-tech, automated birds of prey you can install on your roof that will scare pigeons and other birds away.

- Keep Your Yard & Garden Clean | Birds need something to eat, and making the space around your home uninhabitable for them will make them go elsewhere. Keep your yard and garden clean and ensure there are no food sources around your home including pet food. If you have trash bins around your house, store the rubbish under a lid or in tightly sealed plastic bags.

If you want information on the decline of solar panels, visit the National Renewable Energy Laboratory (NREL), https://www.nrel.gov/docs/fy12osti/51664.pdf. They have done some studies examining the long-term degradation rates of various photovoltaic panels.

The Author's Final Thoughts

Never bend your head. Always hold it high. Look the world straight in the eye.

HELEN KELLER

There are many ways information is available, but it can be exhausting sifting through the unbelievable amount of data. Be aware when your mind needs rest, put on the brakes. When you are ready to start again, take your time, research the information, and ask questions until you are satisfied. Questions ensure the reliability of information you consume keeping your critical thinking skills on point. Realize that when you are not informed, you become vulnerable to fraud and abuse. Therefore, always be well-informed before you make a decision. It will help you recognize high-quality from poor-quality content.

All we can do is make the best decisions we can with the best information we have at that time and place. And learn how to rebound, reinvent, and regroup. Remember—people who seem to move through life with confidence aren't confident about the outcome of a decision; they're confident that they can deal with the outcome, good or bad.

STEPHANIE BOND

This chapter provided you with information on monitoring the production of your solar system, storing electricity, maintenance, being off-grid, and the life of solar panels. It is also the conclusion of Part 1. The next part will start off with answers to some common questions and definitions on words you may not know.

The Resources chapter is divided in 3 parts: domestic,

international, and by country. You will find solar links for Australia, Canada, China, France, Germany, India, Ireland, Italy, Japan, Mexico, Spain, and United Kingdom.

My hope is that *Going Solar The Homeowner's Handbook* provided you with the information you needed to raise your energy awareness and have a better understanding of the different processes of a solar power system. If you want more information, visit my website at www.lourdesdirden.com.

Knowing is not enough; We must apply. Willing is not enough; We must do.

BRUCE LEE

PART 2

You have power over your mind — not outside events.
Realize this, and you will find strength.

MARCUS AURELIUS

CHAPTER 11
FREQUENTLY ASKED QUESTIONS

I never learn anything talking. I only learn things when I ask questions.

LOU HOLTZ

Home

Do I need to notify my insurance company that I plan on installing a solar system?
You may want to ask if they cover any damages to your home during installation.

What happens if I move after I purchase my solar system?

- You may be able to add the cost to the price of the home.

- You can hire a company to transfer the system to your new home.

• If you leased the system, the contract may
be transferred to a new location or to the new owner.

If I sell my home, will the solar system affect the value?
It depends on your solar system's ownership structure. When
the time comes for your home to be appraised, consult with a
real estate professional who is properly trained in evaluating
the impact of solar. If you own the system, it may help boost its
resale value. If you leased the system, you may have a few
options depending upon your lease agreement.

Why is there a lien on my home?
If you have a solar loan and payments have not been made, you
may want to speak with the finance company. Ask if there is a
lien or will a lien be placed on your home for non-payment. If
the solar system is leased, it is owned by a third party. The lien
provides assurance that the homeowner cannot claim the solar
as their own in case of a breach of contract or default on their
payments. For more information, refer to Chapter 8,
"Mechanics Lien Warning."

Are there any requirements if I live in a historic district?
Historic districts must go through an additional step in the
permitting process to ensure the location and method of installation comply with local historic requirements. In some jurisdictions, this review is part of the standard permit review
process, while others are handled by a separate review board.
Ask your solar provider about the requirements in your area.

*Can I purchase a solar system if I am a member of the Homeowners
Association?*
Many states in the United States have passed legislation
preventing these associations from denying members the right
to install solar. Although, they may place restrictions on how

and where solar is installed on your property. Check out state-specific information on their solar laws.

Do solar panels create holes in the roof?
Yes, but installers take precautionary steps to prevent leaks and other damages from occurring. That is why it is important to use a licensed and qualified solar professional.

How much weight can a roof handle?
The weight depends on the type of roof and solar panels. It is critical to have a solar professional evaluate your roof to determine if additional support is required to complete the installation.

Can solar panels ruin your roof?
Solar panels will not damage your roof as long as the roof is in good condition and installed properly by a qualified professional (Chapter 2, "The Roof Matters").

Solar System

Is solar power my only source of electricity?
It depends. If your system is connected to your utility grid, then solar power is not your only source of power. The utility grid ensures a continuous source of power, even when your panels are not producing energy. If your solar panels are off-grid, then you are not connected to the utility grid and solar power is your only source of electricity.

What happens if my panels produce more energy than my home needs?
If you live in a region that allows net energy metering, you may receive solar credits on your utility bill (Chapter 6, "Net Energy Metering"). However, because not all net metering programs are

the same, contact your utility company and find out how much credit you will receive.

How are solar systems measured?
They are measured by the electrical power they produce, in watts (w), kilowatts (kW), or megawatts (mw).

What is the difference between on-grid and off-grid solar power?
On-grid means your solar system is connected to your local utility company's system. Off-grid means you are not connected to your local utility company's grid.

Does a solar system operate at night?
No, they require sunlight to generate electricity for your home.

What happens to my solar system when the power goes out?
Your solar panels will automatically stop producing electricity. This is a required safety feature, designed to prevent your panels from feeding electricity into the grid and injuring the utility lines personnel who are servicing the wires.

Are solar panels recycled?
Yes. The Solar Energy Industries Association has a list of companies that provide this service. For more information, go to https://www.seia.org/initiatives/seia-national-pv-recycling-program.

What is an electricity usage offset?
The percentage of a home's electrical consumption provided by its solar panels. A simple calculation will show how much of a home's electrical usage is offset by its solar panels. Ask your solar provider for more information.

Will solar electricity offset all of my electricity costs?

There are 2 factors that determine your electricity costs and how much you can offset. The first is the ability to produce energy. There are some reasons your solar panels can have less energy production, such as:

- unavailable roof space

- roof may be shaded by trees

- roof may be blocked by vents or skylights

- home has not switched from incandescent light bulbs to LED lights which uses less energy

The second factor is your energy consumption. It will depend on the electronic devices and appliances that you run. Appliances that use a lot of energy include freezers, electric vehicles, inefficient appliances, and heating or cooling systems.

What is solar panel degradation rate?
Over time, solar panels lose their ability to absorb sunlight and convert it into solar energy due to factors such as hotter weather, and the natural reduction in chemical potency within the panel. This is what is referred to as the "degradation rate." The lower the degradation rate, the better the panel. When a solar panel has a lower degradation rate, it will produce more energy over its lifetime.

Should you replace your solar panels after 25 years?
Solar panel systems will keep producing electricity even after the 25-year warranty period if they are well maintained. Although, they may not be as efficient at energy production versus when they were first installed.

Battery

Can I install solar panels and add battery storage later?
Yes, this is called a storage retrofit. For more information, contact your solar provider.

Is a permit required to install a battery?
Not always. Your local permitting office will have requirements for how your battery system should be installed. Ask your solar provider for more information.

How many batteries and space will I need?
The amount of batteries depends on your home and your specific needs. The amount of space depends on how much storage you want. Ask your solar provider for more information.

Net Energy Metering

Why is my true-up bill high?
The following are some reasons for a high bill.

- Your system may not be working as well as it should. The easiest way to prevent problems with your system is to have your solar installer walk you through the process of operating it. This allows you to bypass any potential issues and understand how to read your system's output. If you ever discover any problems with your system, contact the team who installed it immediately.

- Your home is using too much power. For example, running a central air conditioner during a hot summer could be consuming much more energy than normal. Also, old appliances use more energy. It is probably one of the biggest reasons why you are paying more on your electric bill. *Solution: Upgrade your appliances with new energy-efficient models. A new energy-efficient refrigerator uses about 4 times less electricity than an older model.*

- You are reading the meter wrong. Contact your utility company or solar installer for assistance and learn to read your system correctly.

How do I read my electric bill?
Once you install solar panels, your monthly electric bill may look slightly different. Contact your utility company for assistance.

Does my utility company compensate me for the solar energy my system produces? As long as your utility company is part of the net energy metering program, and you generate more power than you use, you will see a credit applied to your bill (Chapter 6, "Net Energy Metering"). However, because not all net metering programs are the same, contact your utility company and find out how much credit you will receive.

Which states in the United States have metering aggregation?
As of 2020, according to the Database of State Incentives for Renewables & Efficiency (DSIRE), the states are Washington, Oregon, California, Nevada, Utah, Colorado, Arkansas, Missouri, Minnesota, West Virginia, Pennsylvania, New York, Maine, Rhode Island, Connecticut, New Jersey, and Delaware. For more information, visit www.dsireusa.org.

Financing

What is the difference between a solar lease and a solar PPA?
The solar PPA is a flexible payment based on how much electricity your solar panels produce. The solar lease is a fixed payment based on the expected electricity production of the solar panels.

What is a hard inquiry on your credit report?
Credit inquiries occur whenever someone accesses your credit account. The credit bureaus note this information, record the date, and the name of the company or entity accessing your information. Hard inquiries refer to instances when a lender accesses your credit report and looks at the information in your report to decide whether to approve or deny your application for credit. For more information, refer to these 2 links: https://www.equifax.com/personal/education/credit/report/understa hard-inquiries-on-your-credit-report/ and https://www.exper ian.com/blogs/ask-experian/what-is-a-hard-inquiry/.

CHAPTER 12

GLOSSARY

The investigation of the meaning of words is the beginning of education.

ANTISTHENES

array
An array is created by combining several solar panels together.

azimuth angle
The direction from where sunlight is coming. At solar noon, the sun is always directly south in the northern hemisphere and directly north in the southern hemisphere.

building-integrated photovoltaics (BIPV)
They are dual purpose and serve as both the outer layer of a structure and generate electricity for on-site use or export to the grid.

contractor
A person or company that undertakes a contract to provide

materials or labor to perform a service or do a job. Contractors run their own business and sell their services to others. They are sometimes called independent contractors.

direct current (DC) & alternating current (AC)
In DC, the electric charge only flows in one direction. In AC, the electric charge changes direction periodically and also periodically reverses because the current changes direction.

electrical panel
A steel box that holds multiple circuit breakers wired to circuits that distribute power throughout your home. It is also called a breaker panel, load center, service panel, and breaker box.

filament
The wire or thread inside the bulb which lights up when you turn on the electric current.

incandescent light bulb
An electric light with a wire filament heated until it glows.

interconnection
The process by which an electric generating facility is allowed to connect to and supply the grid with power. It is an approval process to connect an electrical resource to the electric grid.

inverter
A device that converts direct current (DC) to alternating current (AC).

kilowatt (kW)
One kilowatt is equal to 1,000 watts of electrical power.

kilowatt-hour (kWh)
A unit of energy is equal to 3600 kilojoules. It is used to determine the consumption of electricity, and as a billing unit for
energy delivered to consumers by electric utilities.

LED lighting
LED stands for light emitting diode. LED lighting products
produce light up to 90% more efficiently than incandescent
light bulbs. They are directional light sources, which means
they emit light in a specific direction, unlike incandescent and
compact fluorescent (CFL), which emit light and heat in all
directions.

load
In solar terminology, it refers to the power consumption of a
device that is being used in the system.

main panel upgrade
A home's electrical panel or breaker panel that needs to be
upgraded and replaced.

metering aggregation
It allows a single customer with multiple meters on the same
property or an adjacent property to receive the benefits of net
energy metering.

micro inverter
A plug-and-play device used in photovoltaics that converts
direct current (DC) generated by a single solar module to alternating current (AC).

module
Another word for solar panel.

nameplate capacity
It is the number registered with authorities for classifying the power output of a power station usually expressed in megawatts.

photovoltaic
It means electricity from the energy of the sun. It is derived from the word "photo" with the Greek meaning light and "voltaic" meaning voltage.

photovoltaic cells
Also called PV cells, and they convert sunlight directly into electricity through a process called photovoltaic effect.

photovoltaic shingles
Also called solar shingles, and are solar panels designed to look like and function as conventional roofing materials such as asphalt shingle or slate. These solar shingles produce electricity, and they are a type of solar energy solution known as building-integrated photovoltaics (BIPV).

placed in service
A solar system is considered placed in service once it has been completely installed, and it can be used by the owner. This is one of the main requirements for the Investment Tax Credit (federal tax credit).

renewable energy
Energy that is collected from renewable resources such as sunlight, wind, rain, tides, waves, and geothermal heat.

roofer
Also called roof mechanic and roofing contractor. Roofers

replace, repair, and install the roofing of buildings using a variety of materials.

roof peak
The highest point of a roof, and a protective covering that covers or forms the top of a building.

storage retrofit
This means when you want a battery storage after your solar panels have been installed.

string inverter
A device for converting DC to AC power. The solar panel array is wired in series, rather than in parallel.

subcontractor
A business or person that carries out work for a company as part of a larger project.

utility bill
A detailed monthly invoice issued by a utility company.

utility grid
A commercial electric power distributor system that takes electricity from a generator, transmits it over a certain distance, then takes the electricity down to the consumer through distribution.

watts
Units used to describe power output.

CHAPTER 13
RESOURCES

You have all the tools and resources you need. What you do with them is up to you.

CHERIE CARTER-SCOTT

This chapter is divided in 3 sections: domestic, international, and by country specific for Australia, Canada, China, France, Germany, India, Ireland, Italy, Japan, Mexico, Spain, and United Kingdom.

DOMESTIC

Better Business Bureau (BBB)
https://www.bbb.org
BBB investigates illegal schemes and frauds.

California Public Utilities Commission (CPUC)
https://www.cpuc.ca.gov/aboutus/
CPUC regulates privately owned public utilities in the state of California, and they offer the following two programs: GRID

Alternatives & SGIP. The CPUC presents The California Solar Consumer Protection Guide (https://www.cpuc.ca.gov/ solarguide/) and recommends that solar providers give out this guide during their first contact with potential customers.

GRID Alternatives
https://gridalternatives.org
It is a statewide program for the Single-family Affordable Solar Homes Program (https://gridalternatives.org/what-we-do/ program-administration/sash). They provide upfront rebates to help low-income homeowners access the benefits of solar power, install solar power systems, and provide job training for underserved communities. They operate in the United States, Mexico, Nicaragua, and Nepal.

Self-Generation Incentive Program (SGIP)
https://www.cpuc.ca.gov/sgip/
SGIP offers rebates to California households for a solar battery.

California Distributed Generation (DG) Statistics
https://www.californiadgstats.ca.gov
This is the official public reporting site of the California Solar Initiative (https://www.cpuc.ca.gov/General.aspx?id=6043). They also provide information on incentives.

California Energy Commission
https://www.energy.ca.gov
This is the state's primary energy policy and planning agency. It reduces energy costs and environmental impacts of energy use.

California Natural Resources Agency
https://data.cnra.ca.gov/group/energy
They restore, protect and manage the state's natural, historical

and cultural resources. They publish information on several
topics including energy.

Center for Sustainable Energy (CSE)
https://energycenter.org
CSE is a nonprofit with expertise in renewable energy, clean
transportation, building performance, and energy efficiency.
They partner with state government agencies in Arizona, California, Connecticut, Hawaii, Massachusetts, Nevada, New
Jersey, New York, and Oregon.

Consumer Financial Protection Bureau (CFPB)
https://www.consumerfinance.gov
CFPB protects consumers from unfair, deceptive, or abusive
practices and can take action against companies that break
the law.

Contractors State License Board (CSLB) Department of
Consumer Affairs
https://www.cslb.ca.gov/Consumer.aspx
In California, the CSLB protects consumers by regulating the
construction industry through policies that promote the health,
safety, and general welfare of the public in matters relating to
construction. Complaints against a contractor can be filed with
the CSLB.

Database of State Incentives for Renewables & Efficiency
(DSIRE)
www.dsireusa.org/
DSIRE searches for the latest federal, state, local, or utility
incentives and rebates in the United States. It is operated by the
North Carolina Clean Energy Technology Center at North
Carolina State University (https://nccleantech.ncsu.edu) and
provides up-to-date information about current state incentives.

North Carolina Clean Energy Technology Center (NCCETC)
https://nccleantech.ncsu.edu
The NCCETC is located at the North Carolina State University,
and they provide support for clean energy technologies,
practices, and policies. They also administer the Database of
State Incentives for Renewables & Efficiency (www.
dsireusa.org/), a resource providing financial incentives and
policies.

Federal Trade Commission (FTC)
https://www.ftc.gov
The FTC has a department called the Bureau of Consumer
Protection (https://www.ftc.gov/about-ftc/bureaus-offices/
bureau-consumer-protection), and they protect consumers of
unfair business practices. Complaints help the FTC and other
law enforcement agencies bring scam artists to justice and put
an end to unfair and misleading business practices. The FTC
cooperates with international agencies and organizations to
protect consumers in the global marketplace.

Interstate Renewable Energy Council
https://irecusa.org/about-irec/
This council is a nonprofit that builds the foundation for rapid
adoption of clean energy and energy efficiency to benefit
people, the economy, and our planet.

National Association of Home Builders (NAHB)
https://www.nahb.org/why-nahb/about-nahb
The NAHB of the United States works to achieve professional
success for its members who build communities, create jobs,
and strengthen the economy.

Puerto Rico Electric Power Authority (PREPA)
https://aeepr.com/en-us/Pages/HomePage.aspx

PREPA uses technologies of the electrical industry worldwide, and it provides electrical energy services. All electrical systems in Puerto Rico are under a single entity.

Public Utilities Commission (PUC)
In the United States, the PUC regulates the rates and services of a public utility, such as an electric utility. Their role involves the regulation of essential utility services such as energy, telecommunications, and water. The PUC is in all 50 states and territories. Other names used are *utilities commission, regulatory commission, energy bureau, utility regulatory commission (URC), or public service commission (PSC).* You can find websites for your state's commission by typing the name of your state and "public utilities commission." For example, "California Public Utilities Commission."

Solar Energy Industries Association (SEIA)
https://www.seia.org/initiatives/whats-solar-renewable-energy-credit-srec
The SEIA works with companies and other partners to fight for policies that create jobs in every community.

Solar Renewable Energy Certificates (SRECs)
https://www.energysage.com/solar/cost-benefit/srecs-solar-renewable-energy-certificates/
SRECs are solar incentives that allow homeowners to sell certificates for energy to their utility company. To learn more, visit the SEIA Partner EnergySage, https://www.energysage.com/seia/

Solar United Neighbors
https://www.solarunitedneighbors.org
A non-profit dedicated to representing the needs and interests of solar owners and supporters across the country. They envi-

sion a clean, equitable energy system that directs control and benefits back to local communities.

State Attorneys General
https://www.usa.gov/state-attorney-general
They advise and represent their legislature and state agencies and act as the "People's Lawyer" for the citizens. Select your state to connect to your state attorney general's website. The Consumer Protection Division in the State Attorneys General protects consumer's rights.

State Consumer Protection Offices
https://www.usa.gov/state-consumer
This is your online guide to find your state or territory Consumer Protection Office. They protect consumers from unfair, deceptive, or abusive practices.

United States Department of Energy (DOE)
https://www.energy.gov
The DOE ensures America's security and prosperity by addressing its energy, environmental, and nuclear challenges through transformative science and technology solutions. Energy Saver (https://www.energy.gov/energysaver/energy-saver) is the DOE's consumer resource on saving energy and using renewable energy technologies at home.

Solar-Estimate
https://www.solar-estimate.org
They provide a list of local certified installers by zip code and company name. They have a guide that answers frequently asked questions about benefits, system basics, choosing a system, financial incentives, and warranty and insurance issues. It was developed with funding from the DOE, and it contains a solar panel calculator that allows consumers to estimate how

many solar panels they need for their home and how much they might save.

United States Energy Information Administration (EIA)
https://www.eia.gov
The EIA collects, analyzes, and disseminates independent and impartial energy information. It is the nation's primary source of energy information and, by law, its data, analyses, and forecasts are independent of approval by any other officer or employee of the U.S. government. The EIA provides a map of electric rates to see how your state compares.

Virgin Islands Water and Power Authority
http://www.viwapa.vi/about-us
It is an agency of the Virgin Islands Government which produces and distributes electricity and drinking water to residential and commercial customers in the territory.

Vote Solar
https://votesolar.org
They work to lower solar costs and expand solar access across the United States.

INTERNATIONAL

Clean Energy Solutions Center, Assisting Countries with Clean Energy Policy
https://cleanenergysolutions.org/resources
They provide policy information and assistance. They help governments design and adopt policies and programs that support the deployment of clean energy technologies.

Greentech Media (GTM)
https://www.greentechmedia.com/

They deliver energy news and insights on the technologies, markets, and business models.

International Association of Better Business Bureaus (IABBB)
https://www.bbb.org/local-bbb/international-association-of-better-business-bureaus
The IABBB is the network hub for BBBs in the US, Canada, and Mexico. It investigates illegal schemes and frauds.

International Confederation of Energy Regulators (ICER)
http://www.camput.org/about-camput/international/
The ICER is a member of the International Confederation of Energy Regulators. It improves public and policy maker awareness and understanding of energy regulation around the world.

International Energy Agency (IEA) Photovoltaic Power Systems Programme
https://iea-pvps.org
They conduct joint projects in the application of photovoltaic conversion of solar energy into electricity.

International Renewable Energy Agency (IRENA)
https://www.irena.org/
The IRENA supports countries in their transition to sustainable energy, and serves as the principal platform for international cooperation.

National Association of Regulatory Utility Commissioners (NARUC)
https://www.naruc.org/about-naruc/about-naruc/
The NARUC is a non-profit organization, and it represents the state public service commissions who regulate the utilities that provide essential services such as energy.

National Renewable Energy Laboratory (NREL)
https://www.nrel.gov
The NREL works to discover scientific solutions to the world's energy challenges. They have a tool called PVWatts Calculator (https://pvwatts.nrel.gov) that estimates the energy production and cost of energy of grid-connected photovoltaic (PV) energy systems throughout the world.

North American Board of Certified Energy Practitioners (NABCEP)
https://www.nabcep.org
The NABCEP offers certifications and credentials for skilled professionals, specialists, and those new to working in the areas of photovoltaics, solar heating, and small wind technologies. They have a partnership with Scantron (https://www.scantron.com/test-site-cities/), who enables renewable energy professionals to take Board Certification exams in over 370 cities in 97 countries.

PV Magazine
https://www.pv-magazine.com/about-us/
A monthly trade publication for the international photovoltaics (PV) community. They concentrate on covering the latest solar PV news. Regional news include Australia, China, France, Germany, India, Italy, Latin America, Mexico, Spain, and United States of America.

Renewable Energy and Energy Efficiency Partnership (REEEP)
https://www.reeep.org/programme-people
The REEEP strengthens markets for clean energy services in low- and middle-income countries.

Solar Energy International
https://www.solarenergy.org

They believe in building the global solar energy workforce to stop catastrophic climate change and create a safe and skilled workforce. They work in North America, Middle East, Africa, and Latin America.

BY COUNTRY

Australia

Australian Government
http://www.gov.au
They have a listing of websites for governments in Australia and where you can research information on solar energy.

Clean Energy Council
https://www.cleanenergycouncil.org.au
This council works with Australia's leading renewable energy and energy storage businesses, as well as rooftop solar installers, to further the development of clean energy in Australia.

NSW Government Fair Trading
https://www.fairtrading.nsw.gov.au
Think Smart is a community education initiative of NSW Government Fair Trading aimed at increasing awareness and understanding of consumer rights.

PV Magazine
pv magazine Australia
A monthly trade publication for the international photovoltaics (PV) community. They concentrate on covering the latest solar PV news.

Renew
https://renew.org.au/about-us/

A national not-for-profit organisation that inspires, enables, and advocates for people to live sustainably in their homes and communities.

Smart Energy Council
https://www.smartenergy.org.au/our-story
This council is the peak industry body for the solar, storage, and smart energy management in Australia.

Solar Trust Centre
https://solartrustcentre.com.au/
This is an Australian based, renewable energy, focused on solar energy news, tools, and resources.

Solar Victoria
https://www.solar.vic.gov.au
Helps Victorians take control of their energy bills, solar rebates, and learn how to apply for loans.

Canada

Canadian Association of Members of Public Utility Tribunals (CAMPUT)
http://www.camput.org/about-camput/
CAMPUT is responsible for the regulation of the electric, water, gas, and pipeline utilities in Canada. It improves public utility regulation, and the education and training of commissioners and staff of public utility tribunals.

China

PV Magazine
pv magazine China
A monthly trade publication for the international photovoltaics

(PV) community. They concentrate on covering the latest solar PV news.

France

PV Magazine
pv magazine France
A monthly trade publication for the international photovoltaics (PV) community. They concentrate on covering the latest solar PV news.

Germany

Bundesverband Solarwirtschaft e.V (BSW Solar)
https://www.solarwirtschaft.de
The German Solar Association acts as an informant, consultant, and mediator in the areas of activity between industry, politics, and consumers.

PV Magazine
pv magazine Germany
A monthly trade publication for the international photovoltaics (PV) community. They concentrate on covering the latest solar PV news.

India

PV Magazine
pv magazine India
A monthly trade publication for the international photovoltaics (PV) community. They concentrate on covering the latest solar PV news.

Ireland

Irish Solar Energy Association (ISEA)
https://irishsolarenergy.org
ISEA promotes solar as a leading renewable energy.

Sustainable Energy Authority of Ireland (SEAI)
https://www.seai.ie/grants/
A solar grant is available to all solar systems installed in the
Republic of Ireland. There are terms and conditions to
receiving this grant, and it can be found on the SEAI website.

Italy

PV Magazine
https://www.pv-magazine.com/2020/05/22/italian-homeown
ers-can-now-install-pv-systems-for-free/
A monthly trade publication for the international photovoltaics
(PV) community. They concentrate on covering the latest solar
PV news. On May 22, 2020, Emiliano Bellini published an
article called, "Italian homeowners can now install PV systems
for free."

Japan

Japan Renewable Energy
https://www.jre.co.jp
Their vision is to create a sustainable society through the development of renewable energy.

Mexico

PV Magazine
pv magazine Mexico

A monthly trade publication for the international photovoltaics (PV) community. They concentrate on covering the latest solar PV news.

Spain

PV Magazine
pv magazine Spain
A monthly trade publication for the international photovoltaics (PV) community. They concentrate on covering the latest solar PV news.

United Kingdom

TheSwitch
https://theswitch.co.uk/energy/guides/renewables/solar-power
TheSwitch has an extensive blog relating to renewable energy, technology, and current affairs.

CHAPTER 14
REFERENCES

It always seems impossible until it's done.

NELSON MANDELA

California Public Utilities Commission, California Solar Consumer Protection Guide, *English Solar Consumer Protection Guide* (January 2021), https://www.cpuc.ca.gov/solarguide/.

Center for Sustainable Energy, *Solar for Homeowners*, https://sites.energycenter.org/solar/homeowners.

EnergySage, *Solar 101: how does solar energy work?* Last updated 4/22/2020, https://www.energysage.com/solar/101/.

EnergySage, *SRECs: understanding solar renewable energy credits*, Last updated 7/15/2020, https://www.energysage.com/solar/cost-benefit/srecs-solar-renewable-energy-certificates/.

Richardson, Luke, Posted May 3, 2018, EnergySage, News Feed, *The History of Solar Energy*, https://news.energysage.com/the-history-and-invention-of-solar-panel-technology/.

SEIA Solar Energy Industries Association, Consumer Protection, https://www.seia.org/sites/default/files/inline-files/SEIA-Consumer-Protection-Factsheet-2018-August.pdf.

SEIA Solar Energy Industries Association, *Residential Consumer Guide to Solar Power*, June 2018, https://www.seia.org/sites/default/files/2018-06/SEIA-Consumer-Guide-Solar-Power-v4-2018-June.pdf.

United States Department of Energy, Office of Energy Efficiency & Renewable Energy, *Homeowner's Guide to the Federal Tax Credit for Solar Photovoltaics*, https://www.energy.gov/sites/prod/files/2020/01/f70/Guide%20to%20Federal%20Tax%20Credit%20for%20Residential%20Solar%20PV.pdf.

United States Department of Energy, Energy Efficiency & Renewable Energy, *Own Your Own Power!* A Consumer Guide to Solar Electricity for the Home, https://www1.eere.energy.gov/solar/pdfs/43844.pdf.

United States Department of Energy, Office of Energy Efficiency & Renewable Energy, Solar Energy Technologies Office, *How Does Solar Work?*, https://www.energy.gov/eere/solar/how-does-solar-work.

U.S. Energy Information Administration, Independent Statistics & Analysis, *Solar Explained*, Last reviewed: December 4, 2019, https://www.eia.gov/energyexplained/solar/.

National Association of Home Builders, *Solar Photovoltaic Ready Roof Preparing a Roof for a Future Solar Installation,* https://www. nahb.org/advocacy/public-toolkits/A-Builders-Toolkit-for-Solar/Solar-Photovoltaic-Ready-Roof.

THE AUTHOR, LOURDES

Lourdes Dirden was born in the 1960s when the Internet did not exist and information was found in books and the library. Her love of reading led her to develop her ability to research. Still, it was not until she ran into situations where further information was needed that her research skills increased. Her natural drive to help people evolved to caring deeply about a person's right to be well-informed. Lourdes' commitment to sharing knowledge is what motivated her to write this book.

Lourdes currently works in the solar industry, and she wrote *Going Solar The Homeowner's Handbook* to provide information and resources to empower anyone interested in having a solar system installed.

If you want further information about residential solar power systems or what Lourdes is working on next, visit www. lourdesdirden.com.

THE COVER DESIGNER, RAFFI

Based in Los Angeles, Raffi is an illustrator and story artist. After receiving a BFA in Animation, he worked as a freelance visual artist with small businesses, writers, and filmmakers. With that experience, he discovered the value of having expertise in various production art areas, including graphic design, conceptual development, and illustrative commercial art.

Since 2008, he also gained experience in the animation industry, and he became a founding partner of R.A.T. Studio, an animation/film partnership that produces original content and collaborates with independent and major productions.

In 2014, his years of experience in the visual arts gave him the opportunity to work in "Themed Entertainment." He contracted with several companies, including Walt Disney Imagineering, to work on numerous attractions, shows, and facility designs for theme parks in Asia and Dubai.

In 2019, he extended his creativity to visual storytelling, working as an illustrator of children's books, and other story and illustrative art. To see his artwork, you can visit www.raf-fiantounian.weebly.com.

THE EDITOR, GRETCHEN

Gretchen Pruett is an editor, writer, and historian based out of Jackson, Michigan. Gretchen works with writers, scholars, and cultural institutions from around the nation to produce polished texts and exhibits that educate and foster critical conversations. In her spare time, she enjoys making artwork, restoring her 19th-century home, and skiing when there is snow.

www.pruettwriting.com

INDEX